YOUR KNOWLEDGE HAS VALUE

Muhammed Ernur Akiner

Storm Drainage Design Project for the River Thaw

GRIN Publishing

Bibliographic information published by the German National Library:

The German National Library lists this publication in the National Bibliography;
detailed bibliographic data are available on the Internet at http://dnb.dnb.de .

Imprint:

Copyright © 2015 GRIN Verlag GmbH
Print and binding: Books on Demand GmbH, Norderstedt Germany
ISBN: 978-3-656-87590-1

This book at GRIN:

http://www.grin.com/en/e-book/287101/storm-drainage-design-project-for-the-
river-thaw

GRIN - Your knowledge has value

Since its foundation in 1998, GRIN has specialized in publishing academic texts by students, college teachers and other academics as e-book and printed book. The website www.grin.com is an ideal platform for presenting term papers, final papers, scientific essays, dissertations and specialist books.

Visit us on the internet:

http://www.grin.com/

http://www.facebook.com/grincom

http://www.twitter.com/grin_com

Storm Drainage Design Project for the River Thaw

Muhammed Ernur AKINER*, [1]

[1]Akdeniz University, Vocational School of Technical Sciences, Campus, Antalya, Turkey

Hydrology has been defined as the study of the occurrence, circulation and distribution of water over the world's surface. It covers a vast area of endeavour and is not the exclusive preserve of civil engineers. Engineering hydrology is concerned with the quantitative relationship between rainfall and runoff and, in particular, with the magnitude and time variations of runoff. This is because all water resource schemes require such estimates to be made before design of the relevant structures may proceed (Chow et al., 1988). Examples include reservoir design, Rood alleviation schemes and land drainage. Each of these examples involves different aspects of engineering hydrology, and all involve subsequent hydraulic analysis before safe and economical structures can be constructed (Viessman et al., 1989). The most common use of engineering hydrology is the prediction of design events. This may be considered analogous to the estimation of design loads on structures. Design events do not mimic nature, but are merely a convenient way of designing safe and economical structures for water resources schemes. Civil engineers are principally concerned with the extremes of nature, design events may be either floods or droughts. The design of hydraulic structures will normally require the estimation of a suitable design flood and sometimes a design drought (Bedient and Huber, 1992). A large range of factors control the shape of hydrographs. These include: precipitation type and intensity, catchment shape, catchment gradient, land use and vegetation, soil type, geology etc. A hydrograph is the time-series record of water level, water flow or other hydraulic properties, and can be used to gain insights into the relationships between rivers and aquifers. Typically, a stream hydrograph shows the fluctuations in stream flow through time and is a commonly available dataset routinely measured to support the management of water resources. For a gaining stream, where groundwater is contributing to stream flow, analysis of the stream hydrograph can indicate the magnitude and timing of this contribution (Patra, 2001). Hydrographs are analyzed to find out discharge patterns of a particular drainage basin, Help predict flooding events, therefore influence implementation of flood prevention measures. Storm Hydrographs show the change in discharge caused by a period of rainfall. See Table 1 for the rainfall and river elevation data of River Thaw.

Table 1. Rainfall and river elevation data of River Thaw.

Date	Time	Rainfall (mm)	Date	Time	River Level (m)
26.07.2007	00:00:00	0	26.07.2007	00:00:00	0.33
26.07.2007	01:00:00	0	26.07.2007	01:00:00	0.33
26.07.2007	02:00:00	0	26.07.2007	02:00:00	0.33
26.07.2007	03:00:00	0	26.07.2007	03:00:00	0.326
26.07.2007	04:00:00	0	26.07.2007	04:00:00	0.326
26.07.2007	05:00:00	0.2	26.07.2007	05:00:00	0.33
26.07.2007	06:00:00	0.6	26.07.2007	06:00:00	0.326
26.07.2007	07:00:00	4.2	26.07.2007	07:00:00	0.345
26.07.2007	08:00:00	6	26.07.2007	08:00:00	0.393
26.07.2007	09:00:00	3.4	26.07.2007	09:00:00	0.412
26.07.2007	10:00:00	2.4	26.07.2007	10:00:00	0.431
26.07.2007	11:00:00	3	26.07.2007	11:00:00	0.472
26.07.2007	12:00:00	3.2	26.07.2007	12:00:00	0.532
26.07.2007	13:00:00	2.4	26.07.2007	13:00:00	0.577
26.07.2007	14:00:00	0.2	26.07.2007	14:00:00	0.6
26.07.2007	15:00:00	0.2	26.07.2007	15:00:00	0.626
26.07.2007	16:00:00	0.2	26.07.2007	16:00:00	0.633
26.07.2007	17:00:00	0.2	26.07.2007	17:00:00	0.637
26.07.2007	18:00:00	0	26.07.2007	18:00:00	0.637
26.07.2007	19:00:00	0	26.07.2007	19:00:00	0.641
26.07.2007	20:00:00	0	26.07.2007	20:00:00	0.641
26.07.2007	21:00:00	0	26.07.2007	21:00:00	0.637
26.07.2007	22:00:00	0	26.07.2007	22:00:00	0.637
26.07.2007	23:00:00	0	26.07.2007	23:00:00	0.633

The flood hydrograph is basically a dual plot of river discharge as a line and rainfall as bars over time. The typical shape is shown in Figure 1 and the main components are labelled.

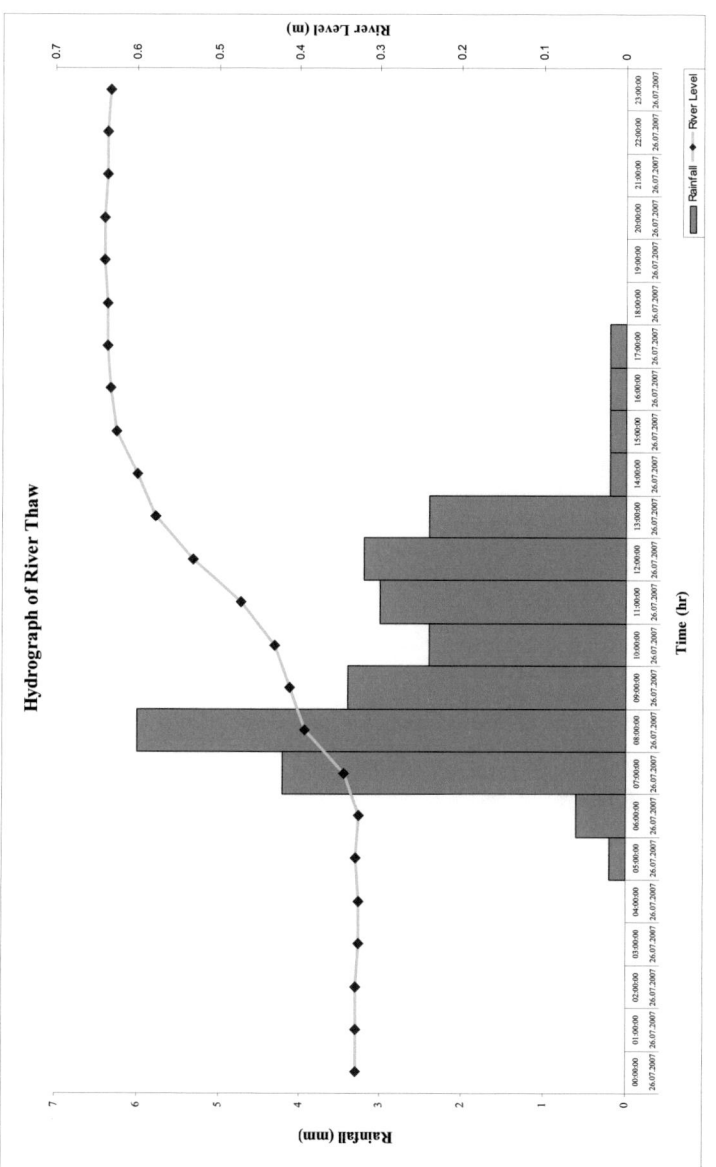

Figure 1. Hydrograph of River Thaw.

Flood lag time shows us the time between the peak rainfall and peak discharge. Peak rainfall is the highest rain amount. Peak discharge is the highest the water level reaches in the river. For our data Flood lag time is approximately 11 hours. Because the peak rainfall is at time 08:00:00 and the peak discharge is at time 19:00:00. A flash flood event is the concatenation of a meteorological event with a particular hydrological situation. To say that a hydrograph is flashy means that the graph depicts sharp vertical jumps and equally steep vertical declines (Rittima, 2008a). What this means for the actual stream represented by the graph is that a flashy stream is one that exhibits significantly increased flows immediately following the onset of a precipitation event and a rapid return to pre-rain conditions shortly after the end of the precipitation. That is to say, water that precipitates within a flashy stream's watershed will make its way quickly from the land into the stream and be flushed through the system rapidly. On the other hand, in watersheds supplying a stream that is not flashy, the transport of water will be slowed through absorption into and seepage through soils, containment on the surface in lakes, and retention in the soil as moisture.

Hydrometeorological factors contribute to flooding. These factors include recent precipitation, soil moisture, snow cover, river ice conditions, stream flow, and forecasted precipitation (Rittima, 2008b). It should be recognized that heavy rainfall is the primary factor that leads to flooding and heavy or excessive rainfall can rapidly cause flooding in any month of the year, even when the flood potential is considered below average. According to Spiegler (1970), quantitative precipitation forecasting (QPF) is a "formidable challenge." Rainfall is a quite ordinary event, which is why it can be difficult to rouse public concern when rainfall becomes life threatening. The public has no difficulty becoming concerned about the threat associated with extraordinary weather events such as tornadoes, but rain is both common and benign in the vast majority of circumstances. Analyses of the impacts on the extreme ends of the flow regime, whether for droughts or floods (Reynard et al., 1998), are still relatively rare. Empirical research into the effects of in-channel and riparian vegetation on flow resistance has been conducted for many years. The effects of vegetation on flow resistance in a range of river environments have largely been unexplored (Chow, 1959). Surfaces covered by nonflexible vegetation are rougher than those covered with flexible riparian vegetation. Based on researches, riparian vegetation minimizes flood risk, while maximizing the environmental benefits of a well developed riparian vegetation cover (Temple, 1982).

Velocity and discharge
Velocity varies vertically and laterally across a river so we refer to mean velocity as velocity. Frequently velocity is measured only at the surface. The length of bed in contact with the water can be measured, and is known as the wetted perimeter. The volume of water passing a point in a given time is the discharge. Table 2 shows the calculated river flow volume (discharge) data of River Thaw. It depends on velocity and cross sectional area at the point. Usually, the equation is written as seen in Equations 1 - 3:

$$Q \quad = \quad A \quad x \quad V \qquad\qquad (1)$$
Discharge (River Flow Volume) = Cross sectional area x Velocity
Velocity is assumed as 3 m/s or 10800 m/hr (See Table 2). In other words;

$$Q = H \times W \times V \qquad (2)$$
Discharge = Water elevation (River Level) x Width x Velocity

$$V = Q \times t \qquad (3)$$
Volume = Discharge x time

Volume of flow versus time curve is depicted (See Figure 2), the area under the curve gives the total drained volume. It is shown in Table 2 with capital v letter. Overland flow is the volume of water reaching the river from surface run off. Through flow is the volume of water reaching the river through the soil and underlying rock layers. Factors influencing storm hydrographs are area, shape, slope, rock type, soil, land use, drainage density, precipitation / temp, tidal conditions, respectively (Gallanagh, 2008).

Table 2. Calculated river flow volume (discharge) data of River Thaw.

Time (t) Hours	Time Interval (Δt) Hours	River Flow Volume (Q) cubic meter / hr	velocity (v) meter / hr	Volume (V) cubic meter	Width (W) meters	Water Level (H) meters	Area (A) square meter
0	1	71280	10800	71280	20	0.33	6.6
1	1	71280	10800	142560	20	0.33	6.6
2	1	71280	10800	213840	20	0.33	6.6
3	1	70416	10800	284256	20	0.326	6.52
4	1	70416	10800	354672	20	0.326	6.52
5	1	71280	10800	425952	20	0.33	6.6
6	1	70416	10800	496368	20	0.326	6.52
7	1	74520	10800	570888	20	0.345	6.9
8	1	84888	10800	655776	20	0.393	7.86
9	1	88992	10800	744768	20	0.412	8.24
10	1	93096	10800	837864	20	0.431	8.62
11	1	101952	10800	939816	20	0.472	9.44
12	1	114912	10800	1054728	20	0.532	10.64
13	1	124632	10800	1179360	20	0.577	11.54
14	1	129600	10800	1308960	20	0.6	12
15	1	135216	10800	1444176	20	0.626	12.52
16	1	136728	10800	1580904	20	0.633	12.66
17	1	137592	10800	1718496	20	0.637	12.74
18	1	137592	10800	1856088	20	0.637	12.74
19	1	138456	10800	1994544	20	0.641	12.82
20	1	138456	10800	2133000	20	0.641	12.82
21	1	137592	10800	2270592	20	0.637	12.74
22	1	137592	10800	2408184	20	0.637	12.74
23	1	136728	10800	2544912	20	0.633	12.66

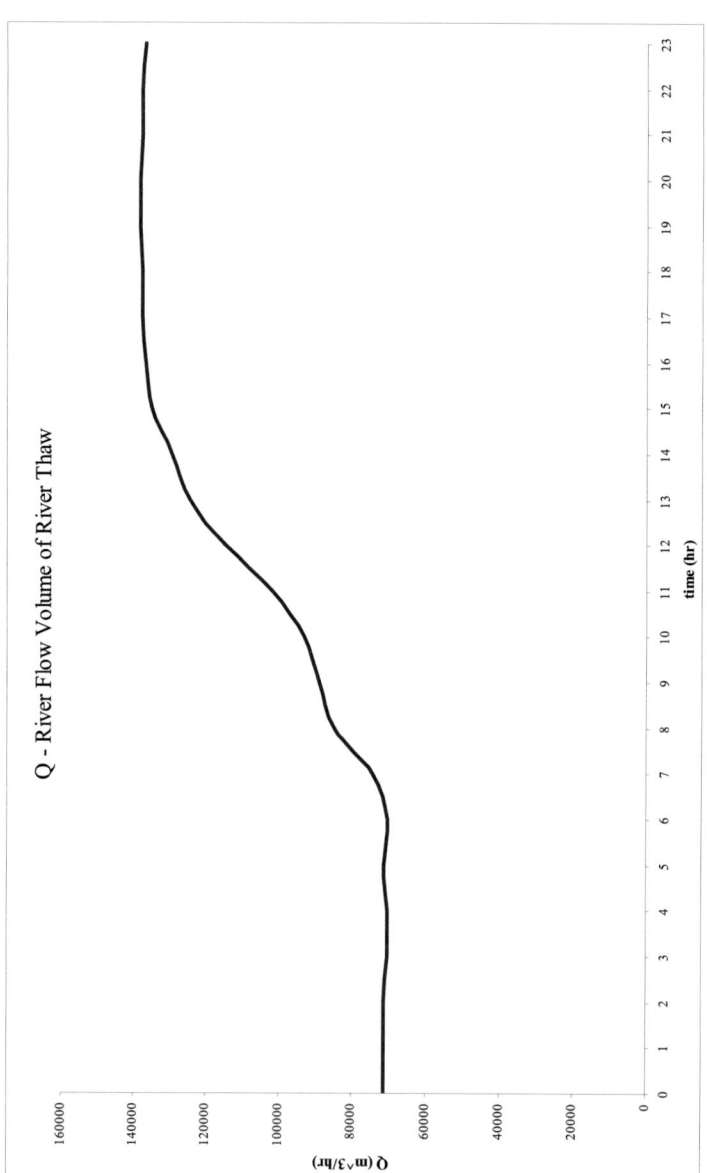

Figure 2. Hourly river flow volume (Q) distribution of River Thaw.

Factors influencing storm hydrographs

The following findings are mainly according as the Fluids and Drainage Engineering module lecture support notes.

Area
Large basins receive more precipitation than small basins, therefore they have larger runoff. Larger size means longer lag time as water has a longer distance to travel to reach the trunk river. At this point the surface area of the Thaw river gains the importance and needs to be investigated in more detail before the possible infrastructure constructions or assembles.

Shape
Elongated basin would produce a lower peak flow and longer lag time than a circular one of the same size.

Slope
Channel flow can be faster down a steep slope therefore steeper rising limb and shorter lag time. Hence, geographical factors are a key factor on the flood risk.

Rock Type
Permeable rocks mean rapid infiltration and little overland flow therefore shallow rising limb.
Soil
Infiltration is generally greater on thick soil, although less porous soils eg. clay act as impermeable layers. The more infiltration occurs the longer the lag time and shallower the rising limb.

River Thaw's geologic development should be clearly identified before installing any water related structure.

Land Use

Urbanization - concrete and tarmac form impermeable surfaces, creating a steep rising limb and shortening the time lag. Afforestation - intercepts the precipitation, creating a shallow rising limb and lengthening the time lag.

Drainage Density
A higher density will allow rapid overland flow.

Precipitation & Temperature
Short intense rainstorms can produce rapid overland flow and steep rising limb. If there have been extreme temperatures, the ground can be hard (either baked or frozen) causing rapid surface run off. Snow on the ground can act as a store producing a long lag time and shallow rising limb. Once a thaw sets in the rising limb will become steep.

Tidal Conditions
High spring tides can block the normal exit for the water, therefore extending the length of time the river basin takes to return to base flow.

Pipeline design considerations
Prevent flooding of carriageway from rainfall and from runoff from adjoining properties. Prevent weakening of sub-grade and pavement caused by excess ground water in cuttings. Prevent erosion of side slopes on embankments and in cuttings.

Channel design considerations
Prevent watercourses from damaging road structure during times of flood. Prevent additional flooding of adjoining properties in immediate vicinity of roadway during times of flood. Skid resistance during heavy rainfall.

References

Chow, V. T., Maidment, D. R., and Mays, L. W. 1988. *Applied Hydrology*. McGraw-Hill Book Co., New York, NY.

Chow, V. T. 1959. *Open-channel hydraulics*. McGraw-Hill, New York.

Temple, D. M. (1982). "Flow retardance of submerged grass channel linings." Trans., ASCE, 25(5), 1300–1303.

Gallanagh M., 2008. Fluids and Drainage Engineering module lecture support notes. Beng (Hons) Civil and Infrastructure Engineering. University of Derby.

Reynard, N. S., Prudhomme, C., and Crooks, S. M.: 1998, Climate Change Impacts for Fluvial Flood Defence, Report to U.K. Ministry of Agriculture, Fisheries and Food, p. 43.

Spiegler D. B., 1970. Snow prediction—A formidable challenge. Weatherwise, 23, 212–220.

Rittima A., 2008a. Lecture Notes EGEN 612 Applied Hydrology. Department of Civil Engineering, Faculty of Engineering, Mahidol University, Bangkok, Thailand.

Rittima A., 2008b. Lecture Notes EGCE 323 Hydrology. Department of Civil Engineering, Faculty of Engineering, Mahidol University, Bangkok, Thailand.

Bedient P. B. and Huber W. C., 1992. *Hydrology and Floodplain Analysis (2nd Ed.)*. Addison & Wesley.

Viessman, Jr., W., Lewis, G.L., Knapp, J. W., 1989. *Introduction to Hydrology*, 3rd edition. Harper and Row, New York.

Patra, K. C., 2001. *Hydrology and Water Resources Engineering*, Alpha Science International.